YOUR KNOWLEDGE HAS VALUE

- We will publish your bachelor's and
  master's thesis, essays and papers

- Your own eBook and book -
  sold worldwide in all relevant shops

- Earn money with each sale

Upload your text at www.GRIN.com
and publish for free

John Tarilanyo Afa

Seasonal and Temperature Variation of soil

GRIN Verlag

Bibliografische Information der Deutschen Nationalbibliothek:

Die Deutsche Bibliothek verzeichnet diese Publikation in der Deutschen National-
bibliografie; detaillierte bibliografische Daten sind im Internet über http://dnb.d-
nb.de/ abrufbar.

Imprint:

Copyright © 2011 GRIN Verlag GmbH
Druck und Bindung: Books on Demand GmbH, Norderstedt Germany
ISBN: 978-3-656-41496-4

This book at GRIN:

http://www.grin.com/en/e-book/213056/seasonal-and-temperature-variation-of-
soil

GRIN - Your knowledge has value

Der GRIN Verlag publiziert seit 1998 wissenschaftliche Arbeiten von Studenten, Hochschullehrern und anderen Akademikern als eBook und gedrucktes Buch. Die Verlagswebsite www.grin.com ist die ideale Plattform zur Veröffentlichung von Hausarbeiten, Abschlussarbeiten, wissenschaftlichen Aufsätzen, Dissertationen und Fachbüchern.

Visit us yon the internet:

http://www.grin.com/

http://www.facebook.com/grincom

http://www.twitter.com/grin_com

JOHN TARILANYO AFA

CURRICULUM DESIGN

TECHNICAL REPORT

STV 678:

SEASONAL AND TEMPERATURE VARIATION OF SOIL

ATLANTIC INTERNATIONAL UNIVERSITY

HONOLULU, HAWAII

2011

Atlantic International University

TABLE OF CONTENT

SEASONAL AND TEMPERATURE VARIATION IN PORT HARCOURT

1.0   INTRODUCTION

The strategic position of Port Harcourt in the Niger Delta region has made the city one of the fastest growing cities in Nigeria. The Niger Delta which is rich in oil and gas provide Nigeria with Ninety-five percent foreign earnings. The oil companies operating in the state has attracted several small, medium and large scale industries. The telecommunication industry which is the fastest growing company in Nigeria has provided job opportunities through several mini companies. Whether small, medium or large companies require the use of electricity and electrical equipments and therefore require proper and safe earthing (grounding). The electrical, electronics and the telecommunication industries require various grounding methods, therefore it is necessary to study the seasonal variation and temperature effect on the grounding systems.

All calculations related to the design of grounding system and the determination of transfer, step and touch potential require information about the soil resistivity at site. Resistivity of soil depends on the physical composition of the soil, moisture content, dissolved salts, season variation, current magnitude and soil temperature. Different soil compositions have different average resistivity but moisture has great influence on the resistivity value of soil. For example, the resistivity of clay varies from 4ohm-m to 150 ohm-m depending on the water content. The study therefore will enable earth system designers, installation engineers to have available records of the earth parameters in this growing city that will enable designers and installer to install safe and reliable system.

2.0   GENERAL ANALYSIS AND BACKGROUND THEORY:

Earthing forms an intrinsic part of the electricity system but it still remains in general a misunderstood subject, even sometimes by well qualified engineers. In recent years there have been rapid developments in the modeling of earthing systems at power frequencies and higher, mainly facilitated by computer hardware and software (Ala and Di-Silvestre, 2009, Lui et al, 2004). This has increased our understanding of the subject at the same time that the design task has become significantly more difficult and emerging standards are requiring a more detailed, safer design. There is thus an opportunity to explain earthing concepts more clearly and a need for this to be conveyed to earthing system designers and installers so that a greater understanding may be gained.

Earth may be defined generally as the reference point in an electrical circuit from which other voltages are measured or a common return path for electric current or a direct physical connection to the earth. Sometimes earth (ground system) can be described as a conducting connection, whether intentional or accidental, by which an electric circuit or equipment is connected to the earth or some conducting body of relatively large extent that serves in place of the earth mass (Cooray et al, 2004,Razevig, 2003).

The most often quoted reasons for having an earthed system are as follows:

- ❖   To provide a sufficiently low impedance to facilitate satisfactory protection operation under fault conditions.

- ❖   To ensure that living beings in the vicinity of substations are not exposed to unsafe potentials under steady state or fault conditions.

- ❖ To retain system voltages within reasonable limit under fault conditions (such as lightning, switching surges or inadvertent contact with higher voltage systems) and ensure that insulation breakdown voltages are not exceeded.

- ❖ To limit the voltage to earth on conductive materials which enclose electrical conductors or equipment.

Others are:

- ❖ To eliminate persistence arcing ground faults.

- ❖ To ensure that a fault which develops between the high and low windings of transformer can be dealt with by primary protection

- ❖ To provide an alternative path for induced current and thereby minimize the electrical "noise" in cables.

- ❖ To provide an exponential platform on which electronics equipment can operate.

- ❖ To stabilize the phase to earth voltages on electricity lines under steady state conditions eg by dissipating electrostatic charges which have built up due to clouds, dust sleet etc.

For these reasons an effective earthing system is a fundamental requirement of any modern structure or system for operational and or safety reasons. Without such a system of a structure, the equipment contained in it and its occupants is compromised.

Earthing systems typically fall into (but not limited to) one of the following categories (Razevig, 2003 Mousa 1994).

- ❖ Power Generation, transmission and distribution

- ❖ Lightning protection

- ❖ Control of undesirable static electricity

❖   Telecommunications

2.1   Power System Earthing

In power system wiring, the ground is to connect with an electrical connection to earth. By connecting the cases of electrical equipment to earth, any insulation failure will result in current flowing to ground that would otherwise energized the case of the equipment. A proper bonding to earth will result in the circuit over current protection operating to de-energize the faulty circuit. A power ground serves to provide a return path for faulty currents and therefore allows a fuse or breaker to disconnect the circuit (Ala et al, 2008, Kenneth et al, 2006).

A particular concern in design of electrical substation is earth potential rise. When very large fault currents are injected into the earth, the area around the point of injection may rise to a high potential with respect to distance points. This is due to the limited finite conductivity of the layer of soil in the earth. The gradient of the voltage may be so high that two points on the ground may be at significantly different potential, creating a hazard to anyone standing on the ground in the area. That is why in substation design the touch voltage, step voltage and the grid potential rise are of great importance (Zeng et al, 2008 Sekioka et al, 2006).

2.2   Lightning Protection:

Lightning protection systems form a very specialized application of grounding used in an attempt to lessen damage to man-made structures and electrical equipment cause by lightning strokes. A lightning protection system is an attempt to provide a preferred low resistance path for the lightning circuit to follow in order to lessen the heating effect of lightning current

flowing through or around flammable structural materials or through porous materials which can contain water such as brick and stone or concrete.

In overhead transmission lines, a ground conductor may also be the top most wire on towers. This ground conductor is intended to protect the power (lines) conductors from lightning strokes. These conductors are connected to earth either through the metal structure of a pole or tower, or by additional ground electrodes installed at regular intervals along the line.

2.3   Control of Static Electricity:

A ground mat is a flat, flexible pad used for working on electrostatic sensitive devices. It is generally made to a conductive metal mesh covered substrate which is electrically attached to ground.

This helps to discharge any static which a worker has built up, as well as any static on tools or exposed components laid on the mat. It is used most commonly on computer repair. Ground are mats also found on fuel trucks, which are otherwise insulated from ground as they make physical contact only with their (rubber and air) tyres. Obviously, static discharge is undesirable during fuel-transfer operations.

2.4   Signal Ground:

An electrical connection to earth can be used as a reference potential for radio frequency signals for certain kinds of antennas. The part directly in contact with the earth (the earth electrode) can be as simple as a metal rod or stake driven into the earth or a connection to buried metals. Because high frequency signals can flow to earth through capacitance, capacitance to ground is an important factor in effectiveness of signal grounds. An ideal signal ground maintains zero voltage regardless of how many electrical current flows into

ground or out of ground. The resistance at the signal frequency of the electrode to earth connection determines its quality.

In television stations, recording studios and other installations where sound quality is critical, a special signal ground known as a technical ground is often installed to prevent ground loops. This is basically the same thing as an a.c power ground but no appliance ground wires are allowed any connection to it, as they may carry electrical interference. In most cases, the studios metal equipment racks are joined together with heavy copper cables (busbars) and similar connections are made to the technical ground (Habjanic and Trlep, 2006, Grcev and Dawalibi, 1990).

To measure the voltage of a single point, a references point must be selected to measure against. This common references point is called ground and considered to have zero voltage. Signal grounds serve as return paths for signals and power (at extra low voltage) within equipment and on signal interconnections between equipment. Many electronic designs feature a single return that acts as a reference for all signals. Power and signal grounds often get connected together, usually through the metal case of the equipment. In some cases signal ground may or may not actually be connected to a power ground.

3.0   MATERIALS AND METHODS

In order to determine the seasonal and temperature variation in Port Harcourt, the city was divided into seven zones. In each of the zones two tests were carried out on the resistivity at the depths of 0.8m, 1.5m, 2.5m and 3.0 meters. These tests were done during the months of February, April, June, September and December. This was an ongoing project but the measurements were concluded in December 2010. The periods (months) reflect the dry seasons, rainy season and the close of the dry and harmattan seasons.

The temperature of the soil was also measured at depths of 0.2m, 0.5m, 0.8m and 1.2m for the same months and at the same areas.

The four point measuring instrument (Wenner Method) was used for the resistivity measurements. Due to the different planting depth the spacing of electrodes also vary. Current terminal was connected to the outer electrodes C_1 and C_2, while the voltage terminals were connected to V_1 and V_2 as shown in fig. 1

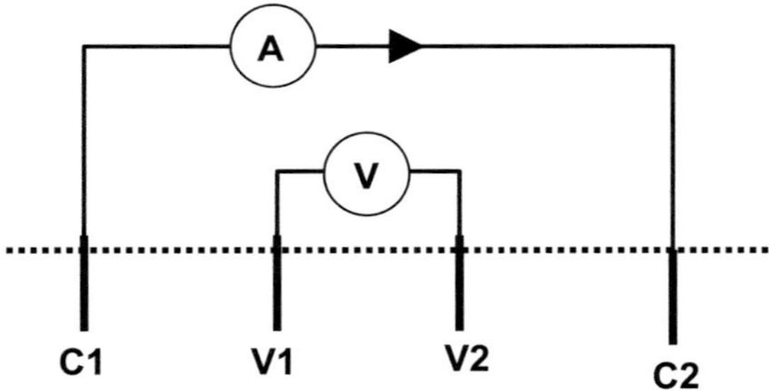

Fig. 1: Four point measuring instrument method (Wenner Method)

The ratio of the measured potential to the calculated current for the given spacing is known as the apparent resistivity value. If b is much less than D (b<<D), the resistivity can be calculated as

$$\rho = 2\Pi DR$$

Where ρ = soil resistivity in ohm-m

      D = Distance between two successful electrodes

      R = the value of V / I_a in ohms

4.0 RESULT

The results of the measurements for soil resistivity for the seven zones are presented in table 1. For the purpose of analysis, (U.S.T/ Diobu) and G.R.A areas were selected. The reason

was that the soil formation is the same with the same topography except in few hilly and depressed areas which is localized to such areas. These results for the analysis are shown in table 2a and table 2b.

Atlantic International University

Table 1: Resistivity Values at Different Depth in Port Harcourt

Month	Depth	Zone 1 Borokiri/Ndoki Site		Zone 2 U.S.T/Diobu Site		Zone 3 Trans-Amadi Ind. Layout Site		Zone 4 GRA Phase1 Site		Zone 5 Mile 4 Site		Zone 6 Chioba Site		Zone 7 Eliozu Site	
		A	B	A	B	A	B	A	B	A	B	A	B	A	B
Feb.	0.8	573	610	672	630	812	750	420	621	740	615	702	660	756	670
	1.5	350	310	402	390	451	432	298	330	400	365	430	402	440	410
	2.5	152	120	130	140	130	125	120	130	150	135	160	140	165	150
	3.0	60	58	45	56	60	61	61	68	70	63	72	65	65	65
April	0.8	440	450	460	400	420	400	360	390	400	400	420	405	390	405
	1.5	220	230	261	225	200	210	200	210	202	251	260	240	211	215
	2.5	102	95	90	98	105	95	87	90	85	92	98	95	90	93
	3.0	52	48	45	50	48	50	52	48	48	52	55	52	48	50
June	0.8	280	320	330	315	340	340	205	300	325	311	350	330	290	311
	1.5	120	130	140	110	130	140	95	125	130	125	108	135	120	120
	2.5	90	95	101	100	95	110	95	98	100	95	91	92	90	95
	3.0	60	52	48	45	48	50	48	40	42	48	52	44	40	45
Sept.	0.8	200	215	200	211	200	190	180	195	190	180	185	200	180	190
	1.5	90	95	100	110	95	91	85	89	85	90	92	92	90	90
	2.5	48	52	45	48	40	42	40	40	40	45	48	50	48	46
	3.0	40	42	38	40	35	36	35	36	34	37	40	41	42	38
Dec.	0.8	524	510	480	500	528	508	320	480	450	480	515	520	490	512
	1.5	260	280	210	220	230	215	210	210	230	220	240	250	220	210
	2.5	110	120	100	101	100	105	105	115	115	105	120	110	112	105
	3.0	50	55	48	48	44	45	48	48	48	45	51	48	46	44

Table 2A: Resistivity value for U.S.T / Diobu Site

Dept. of Elect.	Feb.	April	June	Sept.	Dec.	Coefficient of soil variation
0.8	672	440	380	105	524	6.4
1.5	410	220	120	87	260	4.7
2.5	130	95	95	48	110	2.7
3.0	60	48	46	40	50	1.5

Table 2B: Resistivity value for G. R. A. Site

Dept. of Elect.	Feb.	April	June	Sept.	Dec.	Coefficient of soil variation
0.8	420	360	205	180	320	4.3
1.5	298	200	195	85	210	3.5
2.5	120	87	75	40	105	3.0
3.0	61	52	48	35	48	1.7

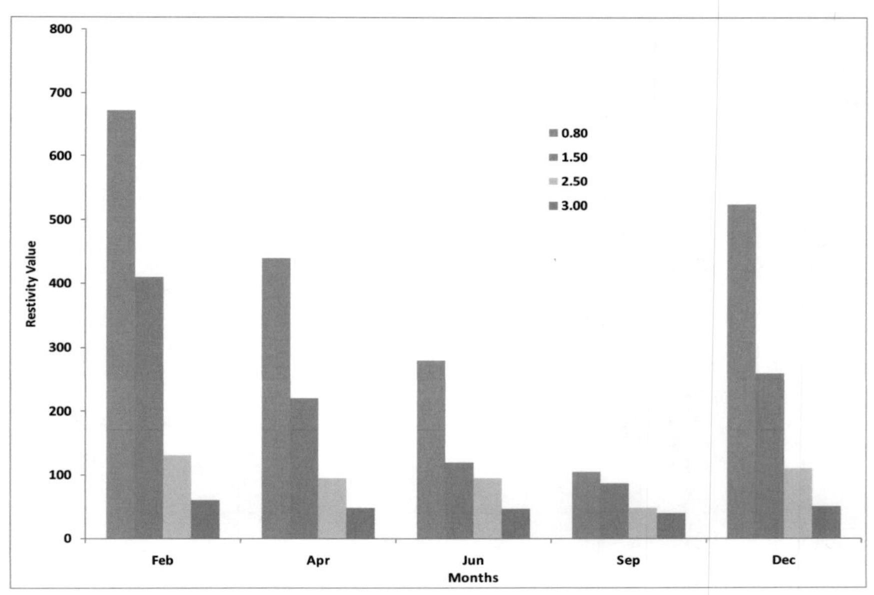

Fig. 2: Resistivity Variation for G.R.A Site

The temperature variation results are shown in table 3

Table 3: Temperature variation of different soil depth

	Temperature in ^{O}C			
Depth in meter	Feb.	Sept.	Feb.	Sept.
0.2	48	27^{0}	42	26
0.5	45	26	39	26
0.8	28	24	28	24
1.2	26	22	25	22
Amb. Temp.	35^{0}	28	33	27
Humidity H_g	65	79	69	79

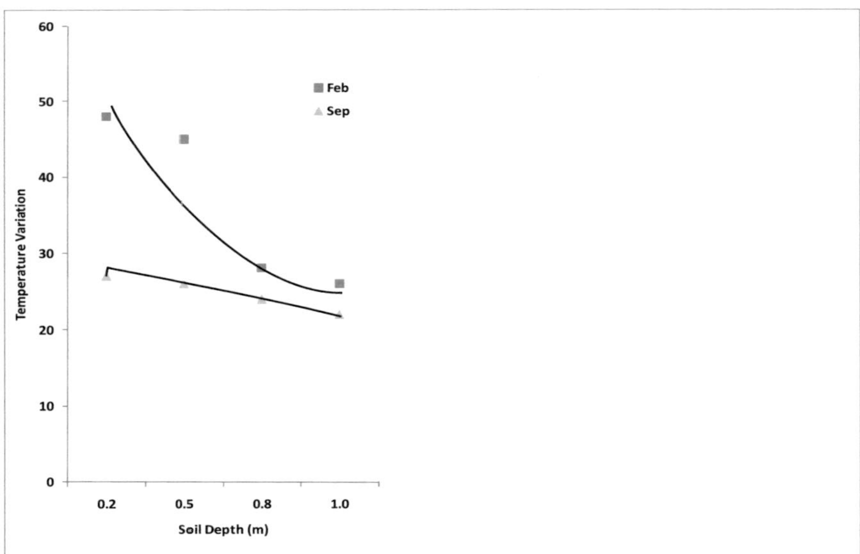

Fig. 3: Temperature Variation for Wet and Dry Seasons.

5.0 DISCUSSION:

The geological and the meteorological factors have considerable influence on soil resistivity.

The dominant soil type is clayey soil and has flat plane in most part of Port Harcourt and its

environs. That is why soil performance in one area can describe soil in several other areas in

Port Harcourt. The only exceptions are some depressed areas and where river canals cut into the land. Such areas are swampy and ponds after prolonged heavy rains. Other areas are those reclaimed swampy areas in some part of the city.

From records in table 2 (a and b), the coefficient of seasonal variation at topsoil was between 4-6 during the dry and the harmattan seasons, but at a depth of 2.5 to 3 meters the soil resistivity was low for every season indicating a permanent moisture level at that depth.

Moisture content of soil has greatest effect on resistivity especially in porous and permeable soil. The resistivity of the soil reduced during the rainy seasons having a coefficient of seasonal variation of about 2-2.5 (Afa, 2010). When rainfalls in the area, some will pond, and some will run off into local surface drainage streams, others may be to natural or artificial drainage, only the remainder will linger long enough to significantly affect electrical resistivity of the soil. From the table of fig. 2, the resistance was lowest during September because of the prolonged period of rain on the land.

In the GRA site, the resistivity of the soil is lower than any other area (site) because of the topography of the area. The soil resistivity reduced drastically after the first rain. The reason was that the soil has a high water acceptance potential, that is, the few months of the rain is enough to make part of the soil saturated especially those areas that are swampy.

Temperature affects the topsoil up to 0.5 meter depth, during the dry and the harmattan periods. This is due to the drying of the topsoil and may even bake due to prolonged hot seasons. At a depth of 0.2 to 0.5 meters the temperature was higher than the ambient temperature. The values are shown in table 3.

At a depth of about 0.8 the temperature effect reduces and become lower than the ambient temperature at a depth of 1.0 meters.

During the rainy seasons the temperature of the soil is always lower than the ambient temperature.

6.0   RECOMMENDATIONS:

Due to the seasonal variation and the permanent moisture table the following recommendations were necessary:

- ❖   For domestic installations, the recommended depth of planting an electrode should not be less than 2 meters. This will enable the effectiveness of the electrode.

- ❖   For all commercial and industrial installation, the electrode depth should not be less than 3 meters except it is counter poise electrode where the contact area is more. Telecommunication tower earthing system may be of lower depth although it may depend on the type of earthing.

- ❖   The accurate measurement of soil resistivity and earthing system resistance is fundamental to electrical safety therefore site measurements must be made to compare with initial results. Periodic measurement of soil resistivity is necessary to ascertain the integrity of the earthing system especially in industrial installations.

7.0   CONCLUSION

Soil resistivity is not only a useful measureable that reflects subsurface structure but also a basic parameters to the design of effective ground and lightning prevention and or protection system. Of several parameters that control soil resistivity (porosity, permeability, mineralization of soil, fraction, ionic content and temperature of pore fluids), only water content and temperature of soil may vary in measurable time scales. The resistivity of the upper layer significantly varied from one point to another, probably reflecting differences in water content in the soil layer due to local topography and drainage.

It is necessary therefore to carryout resistivity tests on project sites before undertaking an installation work .Such local measurement will enable the installer the true situation that will provide a better working data. As much as possible the figures given in the tables are true values of the conditions of the soil both in the dry and wet seasons.

8.0 REFERENCES

1. Abdullahi, M.M., Ali, N.B., Ahmed, KBH, 2010: Elapse time factor on induced vegetative moisture uptake in a saturated soil: Am J. Eng Applied Sci 3: 597 – 603.
2. Afa, J.T., 2010: Subsoil temperature and underground cable distribution in Port Harcourt city: Res. J. of Applied Sciences, Engrg. and Technology: 2(6): 527-531.
3. Ala, G., Di-Silvestre, M.L., 2009: Soil ionization due to High Pulse Transient current leaked by earth electrode: Progress in electromagnetic Res. 4: 1-21.
4. Ala, G., Buccheri, P.L., Romano, P., Viola, F., 2008: Finite difference time domain simulation of earth electrodes soil ionization under Lightning surge condition: IET Science, Measurement and Technology: 2(3) 134 – 145
5. Cooray, v., Zitnik, M., Montano, R., Rahman, M., Liu, Y., 2004: Physical model of surge current characteristics of buried vertical rods in the presence of soil ionization; Journal of Electrostatics: 60(2-4) 193 – 202.
6. Grcev, L. and Dawalibi, F., 1990. An electromagnetic model for transient in grounding systems: IEEE trans. On Power Delivery, 5(4): 1773 – 1781.
7. Habjanic, A., Trlep, M., 2006: The simulation of the soil ionization phenomenon around the grounding system by the finite element method: IEEE Trans. On Magnetics 42(2): 867 – 870.
8. Kenneth, J.N., Jandrell, I.R., Philips, A.J., 2006: A simplified model of the lightning performance of a driven rod earth electrode in multi-layer soil that includes the effect of soil ionization: 41st IAS Annual meeting conference Record of the 2006 IEEE 4: 1821 – 1825.
9. Lui, Y., Theethayi, N., Thottapillil, Gonzales, R.M., Zitnik, M., 2004: An improved model for soil ionization around grounding system and its application to stratified soil: journal of Electrostatics 60(2-4): 203 – 209.
10. Mousa, A.M., The soil ionization gradient associated with discharge of high currents into concentrated electrodes. IEEE Trans. On Power Delivery: 9: 1667 – 1677.
11. Razevig, D.V, 2003: High Voltage Engineering: Khanna Publishers Delhi: 443 – 467.
12. Sekioka, S., Lorentzou, M.J., Phillippakou, M.P., Prousalidis, J.M., 2006: Current dependent grounding resistance model based on energy balance of soil ionization: IEEE Trans. On Power Delivery: 21(1) 194 – 201.
13. Zeng, R., Gong, X., He, J., Zang, B., Gao, Y., 2008: Lightning impulse performance of grounding grids for Substations considering Soil ionization: IEEE Trans. On Power delivery: 23(2) 667 – 675.